# Impressions with Clay

## CREATIVE EXPLORATION INTO THE 21ST CENTURY

Milt Liebson

Schiffer Publishing Ltd

4880 Lower Valley Road, Atglen, PA 19310 USA

# Dedication

To my wonderful family:

My wife, children, grandchildren, and great grandchildren

Library of Congress Cataloging-in-Publication Data

Liebson, Milt.
Impressions with clay : creative exploration into the 21st century / by Milt Liebson.
p. cm.
ISBN 0-7643-2125-0 (pbk.)
1. Pottery craft. 2. Clay. I. Title.

TT920.L53 2005
731.4'2—dc22

2004015488



Designed by Mark David Bowyer
Type set in Korinna BT / Korinna BT

ISBN: 0-7643-2125-0
Printed in China

Published by Schiffer Publishing Ltd.
4880 Lower Valley Road
Atglen, PA 19310
Phone: (610) 593-1777; Fax: (610) 593-2002
E-mail: Info@schifferbooks.com

For the largest selection of fine reference books on this and related subjects, please visit our web site at
**www.schifferbooks.com**
We are always looking for people to write books on new and related subjects. If you have an idea for a book please contact us at the above address.

This book may be purchased from the publisher.
Include $3.95 for shipping.
Please try your bookstore first.
You may write for a free catalog.

In Europe, Schiffer books are distributed by
Bushwood Books
6 Marksbury Ave.
Kew Gardens
Surrey TW9 4JF England
Phone: 44 (0) 20 8392-8585; Fax: 44 (0) 20 8392-9876
E-mail: info@bushwoodbooks.co.uk
Free postage in the U.K., Europe; air mail at cost.

# Contents

# Prologue

Color and design! These are not entities devoid of each other. Rather they combine in a synergistic response to creative expression. I have no hesitation classifying myself as a colorist. I do not create design as a basis for the addition of color. Rather, I enjoy using color to create my design. Each becomes dependent upon each other.

There is lushness to liquid colored clay that I enjoy and its ease of application allows me the freedom of expression that explodes from within to the clay matrix before me. It is not unusual for me to work feverishly for hours, barely looking up, surrounded by pigments and clay, and, at times, splattered with color.

The designs evolve! There seems to be an unending source. As the completed work is produced, a glow of satisfaction at the accomplishment is my reward. If others enjoy this moment through monetary commitment to my work, so much the better.

I pass these thoughts on to you.

Milt Liebson

# List of Gallery Impressions

# Introduction

For many years as an artist my work was devoted to sculpture with wood and stone. Students at several institutions had urged me to put my teaching and technique into book form. This precipitated the publication of *Direct Stone Sculpture*. A second book, *Direct Wood Sculpture* followed duplicating the success of the former. It was obvious to me that the written word has always been necessary. A "how-to" in any field of endeavor has always been popular.

Nevertheless, I began negotiations with a filmmaker to produce a video "Sculpting in Stone." In order to judge the ability of the video producer, I had asked for a sample of his work.

Coincidentally, the sample film was of a potter and print maker from Pennsylvania by the name of Mitch Lyons. He had successfully developed an art form using liquid colored clay. I was overwhelmed with the artistry and beauty of his technique.

This was a new and exciting medium. Utilizing individual creativity, this revolutionary process offers exploration opportunities for both the beginner and accomplished artist. The boundaries of form and design are limitless. The results can be particularly inventive with unpredictable imagery.

Several years of experimentation followed. Early failures gave way to success and eventual personal satisfaction. The need to convey this information to the art-consuming world is at hand.

To the best of my knowledge there is little if any written instructions for this procedure. Are you ready for a process that to this date includes Mitch Lyons and perhaps a handful of professionals beside myself?

Are you ready for a process that does not require a printing press and dangerous chemicals or pigments? Are you ready for a process that requires a minimum investment in tools and materials? Are you ready for exploration into the 21$^{st}$ century?

Welcome to my world of "Impressions With Clay!"

The clay impression is an intriguing hybrid. It is neither a true print nor a painting, but a unique combination of both. Essentially it is a printed painting, using colored liquid clay rather than paint. Each impression is original. Once it is formulated, it cannot be duplicated.

Try this simple test to bring the concept of clay impressions to life. Flatten some clay on a board. Cover it with newsprint and continue to flatten the clay by rolling with a simple roller. Peel off the newsprint. You will notice a thin layer of clay that adheres to the paper.

Imagine the effect if the clay was colored and applied in the form of an artistic design. Imagine the effect if the newsprint was replaced by an adsorbent solid material. This is the basis for the process. This was the original discovery that prompted the development of *Impressions with Clay.*

Color, in the form of colored clay (slip), and individually prepared pastels are applied directly to a clay matrix with a brush, sieve, and or various stencils. An ear syringe can be used to drizzle liquid clay onto a working surface to form a design. The images are then inlaid into the matrix using newsprint and a simple roller. Indirectly, colored slip is applied to newsprint and inlaid into the matrix. There are no boundaries to individual creativity and imagination. Transferring the image from the clay matrix onto interfacing material or wall liner makes the impression. The results are extremely durable and will not fade because the medium is clay, not paint. They can be framed under glass or stretched within a frame.

This is the concept! Now we can proceed with the details of the process.

CHAPTER I

# Materials

All of the materials required can be purchased in stores within your normal shopping areas. Craft shops, lumberyards, hardware stores, paint stores, ceramic supply shops, and art supply shops are the types of dealers that can fill your needs. In addition, local flea markets should not be overlooked.

The following is a generic list of the types of materials required. A detailing of the individual items will appear in subsequent chapters as they are used during the procedures. I have decided to list supplies among the outlets that normally supply them. However, you will find these items available at outlets other than those specified.

Screen door molding.

## Lumber, Hardware & House Ware Supply

**1/2" Exterior Plywood & Screen Door Molding.** Lumber is needed to construct a slab that will serve as a base for a bed of clay to provide a matrix or foundation for the clay impression. The size of the slab can vary with the size of the prints that you will create. Obtain a piece of exterior grade plywood. The dimensions can be 1/2" x 20" x 24". The size can be any dimension you wish, however, the 20" x 24" that I recommend is suitable to start.

You will need about 8 feet of screen door molding. This is required to create a frame around the plywood slab.

**Staple Gun & Hammer.** Do not attempt to use a simple paper stapler. You will need a power stapler for wood, and you may have these items among your current tools. Use 1/2" staples to fasten the molding to the plywood. The hammer is needed to drive the staples flush against the molding.

Staple gun and hammer.

**Electric Blender.** I'm certain that an electric blender is currently part of your kitchen appliance inventory. Once it is used for preparing the liquid clay, you will not want to return it to the kitchen. It is best to obtain one at a local flea market or garage sale. Perhaps you can purchase a low priced, one speed new appliance.

Electric blender.

**White Exterior Paint Sealer.** Two coats of the sealer will be used to prevent the clay from drying too rapidly after it is applied to the plywood slab. You can use varnish or shellac. Paint sealer is preferred.

**Wire Strainer.** This is an ordinary kitchen item available at any house ware department. It will be used to apply pastel colors to the design.

Wire strainer.

**1" x 3" Lumber.** Obtain a piece of prime lumber, 1" x 3" x 24" at any lumber yard. Be certain that it is not warped or defected in any way. It will be used to create the clay slab.

**Dry Wall Tape.** Rolls of dry wall tape are available at the building department of your lumber or hardware dealer. Normally, it is used in construction to seam walls as they are erected. You will use it to create borders around your clay impression.

**Plastic Bags.** These are needed to bag your clay matrix board during times that you leave your studio. The clay matrix cannot be allowed to dry out. I prefer bags of about 36" x 60". Smaller bags can be used as long as they can accommodate the size of your matrix.

**Storage Jars.** Obtain wide-mouth jars for storage of the pigmented liquid clay. Ten or twelve would be adequate for the variety of colors you will use. I use a Rubbermaid product that has a capacity of 473 ml. It has a screw-on cap that provides an airtight seal. Similar tightly sealed glass jars are suitable.

Plastic and glass storage jars.

**Brayers and Scrapers.** Within this group, anything goes! Use your imagination! Browse through house ware, kitchen utensil, and hardware departments. Any gadget that could produce an exotic texture to surface will do.

Brayers and scrapers.

## Art and Art & Craft Supply Stores

**Newsprint.** This product may be new to you. Newsprint is the raw paper prior to the printing of the newspaper. It is available in both smooth and rough grades. I prefer the smooth type. You will find it at art supply dealers in rolls and sheets. Mail order art supply dealers offer large quantities of sheets of newsprint at very reasonable prices. A local newspaper publisher may be willing to sell you the ends of the rolls of their newsprint used in printing. Try them as an inexpensive source of supply. Do not attempt to use newspaper instead of newsprint. I have discovered that the ink of the printed paper will transfer along with the clay.

**Disposable Vinyl Gloves.** You may wish to avoid collecting pigments on your hands. Using inexpensive disposable gloves will do the job. This is particularly important when using organic pigments.

Hoki-style brush.

**Hoki Style Brush.** The Hoki Brush is a soft brush, preferably two-and-one-half to three inches wide. It is used to apply slip to the clay matrix. The professional types are Japanese in origin and are far superior to ordinary wide paint brushes. There are however, many soft wide brushes that can be substituted.

**Assorted Artist Brushes.** You will need six or seven brushes of various sizes. Mop style is preferable for applying liquid colored clay. Try different types to find those of your choice.

Assorted brushes.

**Assorted Stencils.** I use stencils to create patterns and designs. Scraping prepared pastels over a simple kitchen strainer will produce exotic results. Anything goes! Use your imagination! Find them anywhere! The list includes: plastic screens, paper doilies, leaves from your garden, cheerios cereal, cookie forms, plastic tubes, and paper cutouts among others.

Assorted stencils.

## Ceramic Supply Dealers

**Clay Cutter.** The clay cutter is a simple instrument. It consists of a thin stainless steel wire or braided nylon cord stretched between poles. Slices of clay are easily cut for application to the wood slab.

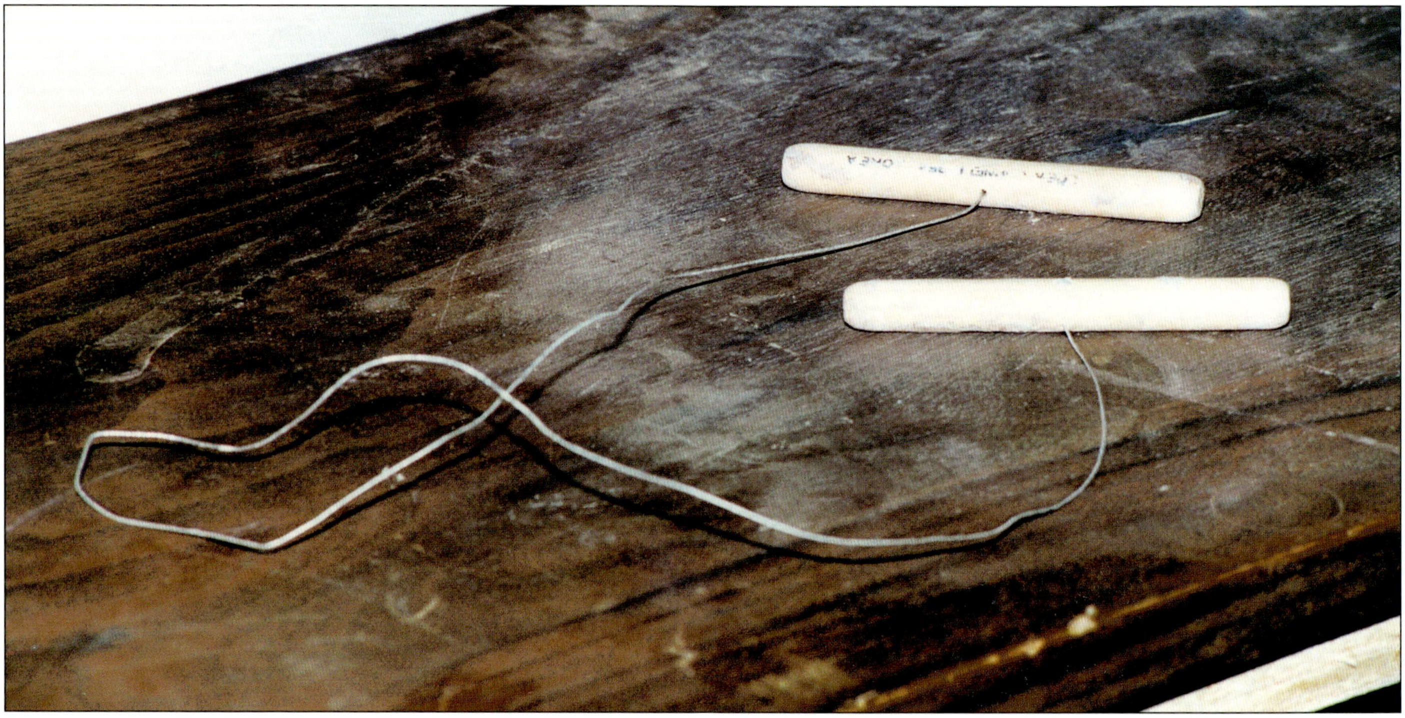

Clay cutter.

**Stoneware Clay.** Stoneware clay is most desired for creating the matrix. This is the base for your inlaid design. It dries evenly and stands up well to periods of inactivity. The clay is inexpensive and you will find it advantageous to purchase fifty pounds packed in two containers of twenty five pounds each. It is available in gray or brown and it does not matter which you choose.

**Kaolin (China Clay).** Kaolin is an inexpensive powdered clay which produces the liquid clay (slip) when mixed with water. You will be using this frequently to prepare colored clay, for coating the matrix, and as a base for pigments. It is best to obtain a twenty-five pound tub of white domestic or English variety.

**Double Pony Roller.** This is the tool with which you will inlay your color and pull the impression. You will need a four-inch and a two-inch roller. Additional rollers may be obtained. These can be of hard rubber, wood, Lucite, and stainless steel. The latter is far the best.

Double pony roller and assorted single rollers.

**Fettling Knife.** A simple, inexpensive stainless steel knife used for trimming clay. These are usually four to five inches long and set in a smooth wooden handle. It is a handy tool. Do not ignore it.

Fettling knife.

**Texture Brushes and Wire Scratch Brush.** The stainless steel brush bristles provide the right degree of stiffness for applying texture. Similarly, heavy textured brushes can produce special effects.

Texture brushes and wire scratch brush.

---

**Water Spray Bottle.** This is simply a plastic bottle with a sprayer top. Get one that holds about thirty-two ounces. You can obtain these almost anywhere in the market place. Garden supply, house wares art supply, and craft shops all supply them. This is a vital item and is used extensively throughout the procedure.

I have left the remaining two items for last. These are the **fabric base** and the **pigments**. All of the aforementioned materials are essential and you cannot proceed without them, especially the plywood slab, the stoneware clay, and the kaolin. But when we think of actual production, it is quite natural to concern ourselves with pigments and the fabric on which to produce our finished product. Each would be considered to be, as it were, a bit out of the ordinary.

This is most evident in the choice of the fabric material upon which the impression is produced.

After experimenting with various papers and other media, and their relative qualities of absorbency, I have found the best to be a nonwoven canvas. This is not to be confused with artist's canvas.

There are two choices. The first is **extra heavy-duty interfacing**. This is the same material that dressmakers use to line and give support to fabrics. Look for extra heavy stabilizer manufactured by the Pellon Company at your local fabric outlet. It is supplied in rolls of 24 to 36 inches and you can purchase as many yards in length as you wish at a modest price.

A better choice, and one that I prefer, can be found at a store that sells wallpaper and paints. Ask for **wall liner**. Normally this product is used to line irregular walls prior to the installation of wallpaper. It is stiffer than interfacing, just as absorbent, and easier to work. Furthermore, it is cheaper.

Pigments are a more normal concern for artists. However, it must be pointed out that artist's oils, acrylics, and watercolors cannot be used. We will be dealing with an aqueous mixture of colored clay to be inlaid onto a clay matrix. Artist's oils will not mix with aqueous media, acrylics have a drying time which would interfere with the process, and watercolors are too thin.

Pigments can be divided into two classes, organic and inorganic. The organic colors are preferred. They are brilliant, vibrant, and spectacular, but they have a downside: they can be toxic. Care should be taken to avoid breathing organic powdered pigment while mixing it with water. Use a dust mask or respirator, or avoid the problem, as I do by using pre-mixed organic colors obtained at local paint supply stores. They are called Universal Paint Pigment Concentrates. These colors are available in various sizes up to sixteen ounces. Prolonged contact with the skin should be avoided. Vinyl disposable gloves are necessary.

Organic pigments and tempera colors.

An excellent alternative is the use of Tempera Paints. You can buy them in dry or premixed form at your local art or craft store. They are nontoxic and produce quality results. I use both.

Universal pigments are available in primary basic colors: blue, yellow, and several shades of red. With proper mixing, you can prepare green, purple, orange, and other shades between. Tempera paints are available in a wide range of colors.

Inorganic colors are earth toned and lack the brilliance and intensity of organic colors. These are the potter's colors available at ceramic supply stores, masonry, or building supply stores. I use organic pigments and Tempera colors exclusively.

The following is a list of pigments that I use. These were used to produce the works in the Gallery of Clay Impressions further in this text.

| Organic Universal | Tempera Liquid | Tempera Dry |
|---|---|---|
| Thalo Blue | Orange | Peach |
| Bulletin Red | Black | Magenta |
| Yellow, Light | Brown | Blue |
| Permanent Red | Red | Brown |
| Orange | | |
| Red-Orange | | |

### And a Word about Workspace

A proper workspace is vital. This procedure cannot be performed on upright easels or small tables. You should have several sturdy flat tables that are 2-3 feet wide and about 4 feet long. Cabinets or shelving to store supplies and pigments are equally important.

My own studio is a building adjacent to my home and provides working space of about 20 ft x 25 ft. The center of the floor is dedicated to sculpture with proper work tables. Along two walls are work counters measuring 2 ft x 16 ft. Above and adjacent to the counters is storage space. These provide ample space for "clay impressions."

CHAPTER II

# Let's Get Started

The formalities are over! We've discussed history, exploration, materials, and sources of supply for our project. Now lets do it!

It would be rather simple for me to list the individual steps in the production of a clay impression. How boring that would be. I've got a much better plan. Lets go into my studio and have one of my students produce an impression under my guidance. We will use all of the techniques that I use and enjoy.

Perhaps we can stimulate the discovery of new techniques through creative explorations. The means by which we advance is discovery and innovation. Regardless of the results, in the final analysis, dynamic advancement is far better than static perfection.

Let me introduce my student to you. Lila is artistically creative. However she has never produced a clay impression. I will give instructions for the procedure and she will actually do the work. Imagine yourself in her place. Lets work together.

The basis for the entire operation is the construction of the plywood slab. This will hold the clay matrix for the inlay of colored clay and the pulling of the final product.

Decide upon the size of the impression that you wish to produce. It can be any size up to the dimension of the available fabric. In this case it would be the dimension of the Pellon Stabilizer or wall liner. I would suggest that for a start, a good size is 20" x 24". As you become more experienced and successful, you can experiment with various sizes. Remember that for each size, a new slab must be constructed.

## Construction of the Slab

From your lumber yard, obtain a piece of 1/2" thick exterior grade plywood measuring 20" x 24". Cover one surface of the plywood with two coats of exterior paint sealer. Allow it to dry thoroughly between coats. You can use waterproof shellac or polyurethane varnish as an alternative.

Plywood slab.

Cut the screen door molding to frame the flat surface of the slab. You will need two pieces 24" long and two pieces 19" long.

Cut the molding to size.

Fasten the molding to the outside edges of the slab to create a frame. Use the staple gun and 1/2" staples. Drive the staples through the striations in the molding.

Coat the slab with exterior paint sealer.

Frame the plywood.

Use the hammer to drive the staples completely within the molding striations. The staples must not protrude above the molding.

Drive the staples.

The completed frame. It will be filled with stoneware clay.

You have created a frame on the slab with a 1/4" void.

## Filling the Void with Stoneware Clay

Use the wire clay cutter to cut slices of stoneware clay about 1/4" thick.

Slicing the clay.

Place the slices about 1/2" apart within the void of the slab until it is loosely filled.

Placing the clay within the void.

Use the heel of your hand to flatten the clay within the void.

Filling the void.

Use a piece of lumber 1" x 3" x 24" to scrape the clay to smooth the surface. Start at the middle. Place the board so that it sits astride the molding and scrape in each direction until the clay becomes flat. It does not have to be smooth, but it must be flat.

Flatten the clay.

Use your fettling knife to scrape unwanted clay from the board.

Use your fettling knife to scrape unwanted clay from the board.

Use pieces of clay to fill any remaining holes and re-scrape until it is flat.

The filled void is now l/4" throughout.

Allow to air dry at room temperature until the surface of the clay is leather hard. This should take from four to eight hours.

Air drying. Progress.

Do not allow the clay matrix to over dry and crack. If you must leave your work area for long periods, place the entire clay slab in a plastic bag and seal with a tie. If you find that the clay surface has developed cracks or small fissures, simply spray the surface lightly with water and scrape till flat.

Bag the clay slab.

This is not to be construed as rest time. There is a lot of preparation for you to complete. The key ingredient for all of your endeavors is our liquid clay, a mixture of kaolin (China clay) and water. We will refer to it as slip. You will be continuously preparing and using it.

The slip has many uses:

• It is used to coat the clay matrix slab to form a base for the inlays.

• It is the base for creating liquid colored clay.

• It is used to prepare the pastels used in the inlay.

• It creates pieces of semi solid and solid colored clay to be used in your design.

**Preparing Slip**

Place about a pint of water in the blender. Add an equal amount of kaolin.

Add water to the blender.

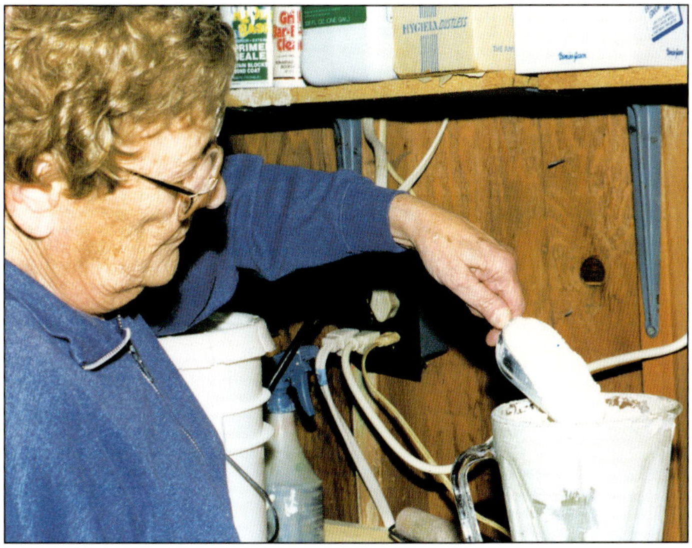

Add kaolin.

Mix at slow, then at high speed for a minute or two until a smooth creamy mixture is obtained. Scrape any powder off the sides and continue to blend until smooth.

The slip should be thick, but flowing. The thicker, the better. If necessary, add more kaolin and continue to blend.

Place the liquid slip in a covered container for future use. A clean three-pound margarine container will suffice. Keep the container cover in place so that the slip does not dry.

Pouring thickened slip.

# Preparing Liquid Colored Clay

## From Dry Pigments

Always wear a dust mask when mixing dry pigments. For safety, avoid inhaling the dust.

Place about two to three teaspoonfuls of dry pigment in a wide-mouth glass or plastic container. Stir in a little water to make a thick paste. Continue stirring and add prepared liquid slip to reach the desired color.

Add dry pigment to jar.

A little water.

Mix well.

Add slip and mix.

## From Liquid Pigments

Half fill a wide-mouth jar with slip. While stirring, add small amounts of liquid Universal Pigment concentrate or tempera liquid to the jar, until the desired color is reached.

Repeat for all the colors you need. Each color should have its own jar.

Begin with the primary colors, red, yellow, and blue. From there you can make every other color by mixing. Pour some of the prepared blue liquid clay and the yellow into a third jar. Mix well to form green. Do the same for red and blue to form purple. If you are not familiar with color blending, refer to any artist's primary color blending chart. Experiment with mixing to obtain a range of colored liquid clay. Keep each jar well closed to prevent drying.

Wash hands thoroughly after mixing colors.

Add liquid pigment concentrate.

Add slip.

Mix well.

# Preparing Solid Pastels and Semi-Solid Colored Clay

Pour some of the prepared liquid colored clay on to a sheet of clean newsprint. Use enough to create a circle of about three or four inches. Pour a variety of colors.

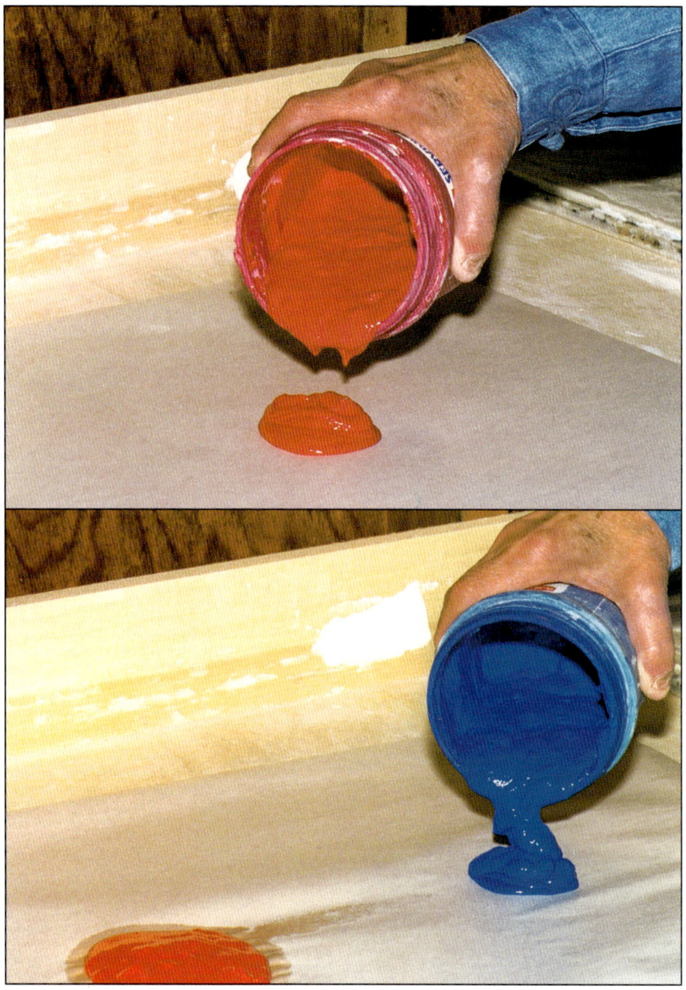

Preparing pastels.

The newsprint will extract the water from the liquid clay, and a circle of wetness will form around the area. Allow the clay to dry until semisolid. Look for the shine to leave the surface of the clay. This should take about six hours. When the clay can be separated from the newsprint it is time to proceed.

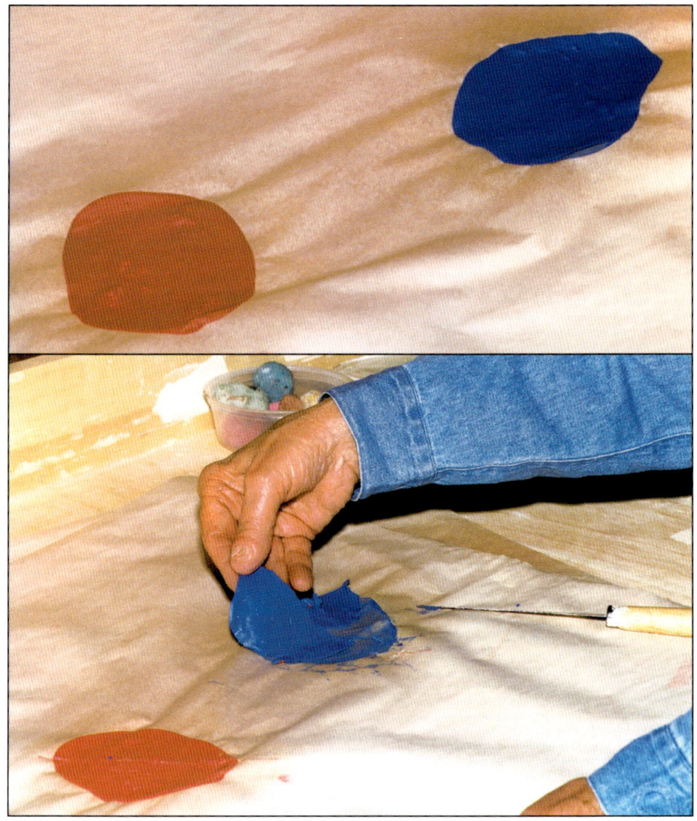

Newsprint will extract water.

Use your fettling knife to slice the circle in two. It is advisable to wear vinyl disposable gloves when handling the moist colored clay. Roll one half into a cylinder about 3/4" wide. Place it on another sheet of newsprint and allow drying for a day or two. The newsprint pulls the water from the clay. Your pastel is now formed.

Cut in two.

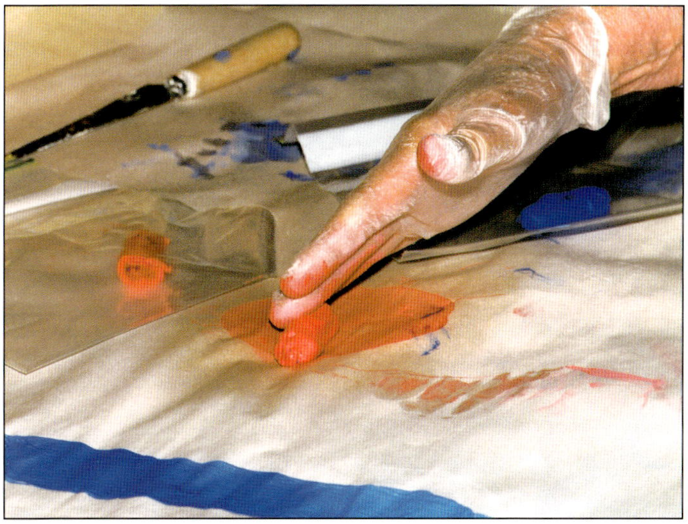

Roll the pastel.

Roll the other half into a cylinder while it is still pliable. Place it into a small baggy and seal it. This part will remain semisolid and can be inlaid into your design.

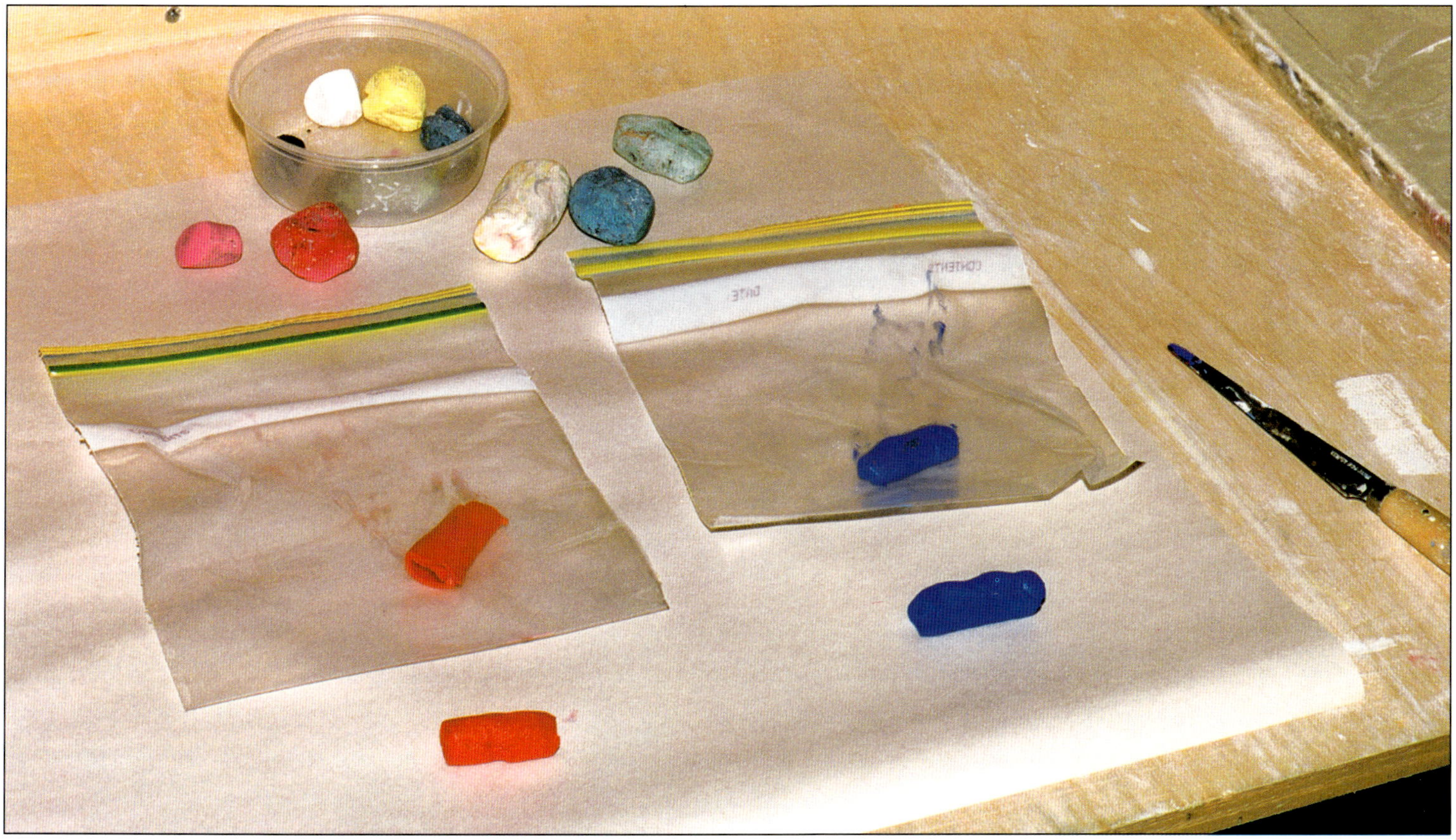

Semi-solids.

Flakes and chips of solid colored clay are useful in creating designs. Use a small brush to thickly apply clay to a sheet of newsprint. Allow the newsprint to extract the water from the clay and allow drying. Chips and flakes of dry colored clay will form. Save these for future applications.

Flakes and chips.

Let us return to the slab and the drying clay matrix. Feel it with your fingers. If it is completely dry, you will find some cracking of the surface. The reason for sealing the board at the beginning of this process was to prevent escape of moisture through the plywood. If the surface is cracked, cover it with a wet towel and bag the entire board for several hours. A proper consistency will soon be established. Start drying again.

It has been drying for at least four to eight hours. Depending upon the warmth and dryness of the surrounding area, the surface should feel "leather hard." The clay under the surface will be pliable. We are now ready to proceed with the completion of the matrix.

### Coating the Matrix with Slip

Use white slip void of tint for this operation. If you need additional slip, prepare another batch using equal parts of kaolin and water. Mix in the blender as before. With a wide Hoki brush, apply an even coat of the slip to the matrix. Allow it to dry thoroughly. This should take about twenty minutes and can be hurried with the use of a fan or hair dryer.

Repeat the process until you have applied five coats of slip, drying well between coats. The final coat should be smooth and produce an even surface. It should be dry to the touch and not sticky. Remember that the slab should not be allowed to dry excessively. If you plan to leave your work for extended periods of time, bag the slab in order to retain moisture.

We are ready for the application of our colors and design.

Coating the Matrix with Slip.

CHAPTER III

# The Inlay of Color and Design

It is the creative explorations of the individual that produces artistic variations. Remember that a design is not based upon a set of rules and regulations. There is no set pattern. If one hundred impressions were prepared, there would be one hundred variations.

Allow yourself to drift into the world of fantasy and make believe. Become an "Alice in Wonderland." You will enjoy the voyage. The exhilaration of success awaits you as you reluctantly return to the world of reality.

The workbench is ready. We have all of the materials and preparations at hand. We are going to work with a 20" x 24" clay slab that has been coated with five coats of white slip.

Examine the edges of the slab where the clay meets the molding. If there is a slight separation, this indicates a sign of readiness. In addition, make sure the surface is dry. If it is somewhat sticky, apply some hot air with a hair dryer to remedy the situation.

Examine the surface of the slip-coated matrix. It should provide a good base for color application.

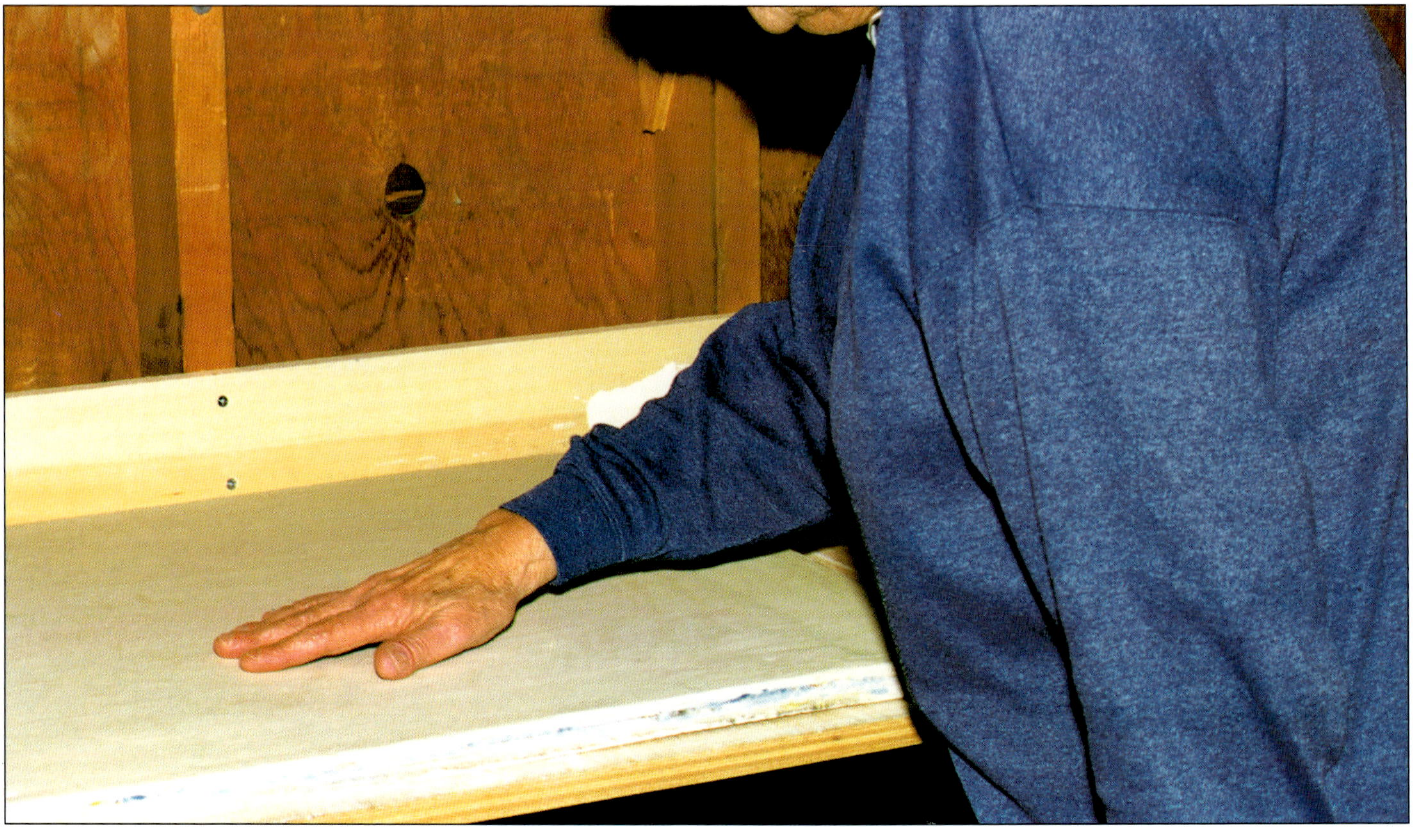

Examine the surface.

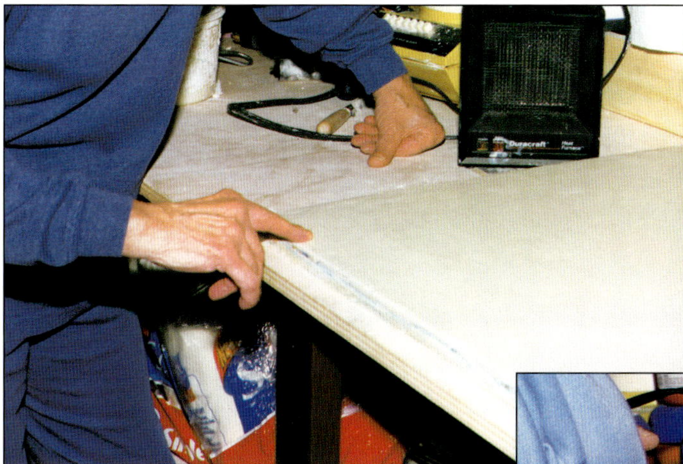

A slight separation indicates readiness.

Use a hair dryer.

There are many ways to inlay color design including:

• Direct application with a brush to the matrix.

• Application with a brush to newsprint, inverting to the matrix, and transferring the design with a roller.

• Drizzle thin colored clay onto newsprint with an ear syringe. Allow it to dry until the shine dissipates then invert it onto matrix. Inlay with a roller.

• Scrape dry pastel through a strainer over a stencil onto the matrix. Spray with water to set, then inlay using newsprint and a roller.

• Place pastel flakes or semisolid pieces of colored clay onto the matrix. Spray with a little water. Cover with newsprint and inlay with a roller.

• Dip end of a tube into liquid colored clay. Press end onto the matrix. Inlay with the aid of newsprint.

• Texture the surface of the matrix after colored clay has been applied. Various brayers and utensils are employed.

I have stated that variations exist. Once you have mastered the technique, you will be free to instill your own design.

Lets start by coating the lower third of the slab with a deep blue colored clay. This represents the direct application technique. Apply the clay liberally with a wide brush from one edge to the other. A thin coat will not do.

Allow the textured blue area to dry until the shine disappears. The design is then inlaid into the clay slab by covering the area with a clean piece of newsprint and rolling with a firm pressure. It is important that the design dries sufficiently before the inlay proceeds. If the area is too moist, it will smear under the newsprint when rolled. If it is too dry, a slight spray with water will soften the clay and allow it to inlay into the matrix.

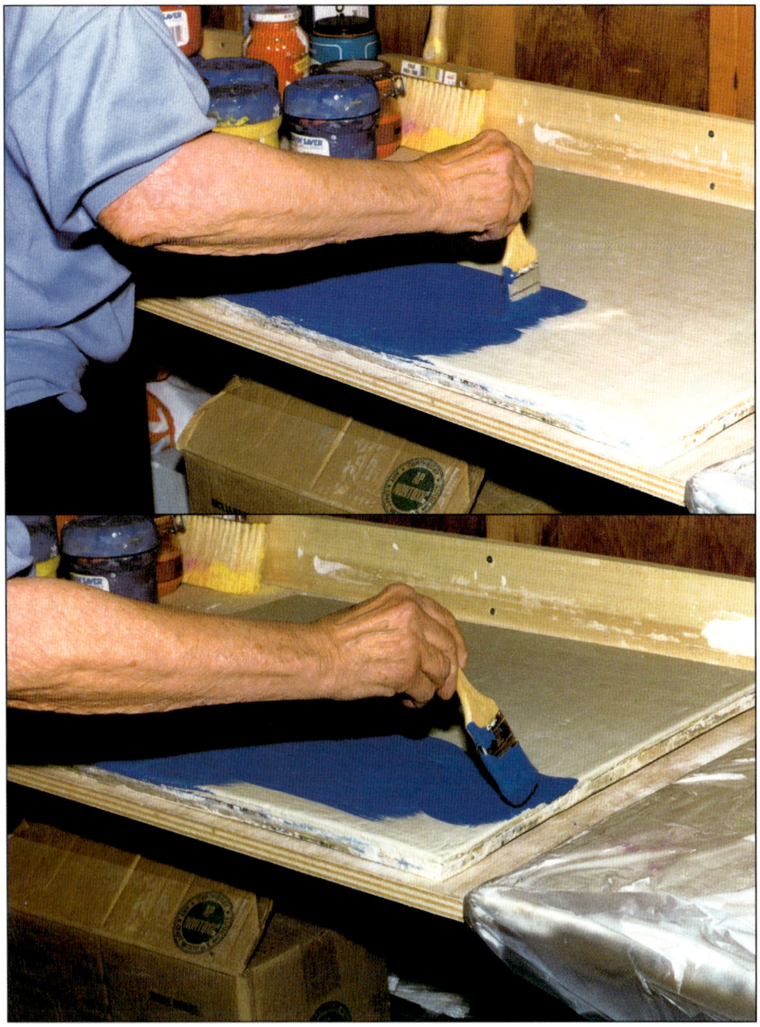

Direct application to matrix.

Inlay into the matrix with newsprint and a roller.

Use a scraper with a distinct serrated edge as shown in the accompanying photograph. Draw the tool through the blue area with a sustained wavy, undulating motion that will remove  blue clay and reveal the white undercoating. This will instill our primary design onto the slab.

Carefully remove the newsprint revealing the inlaid design that we have created.

Use a scraper to create a design.

Carefully remove the newsprint.

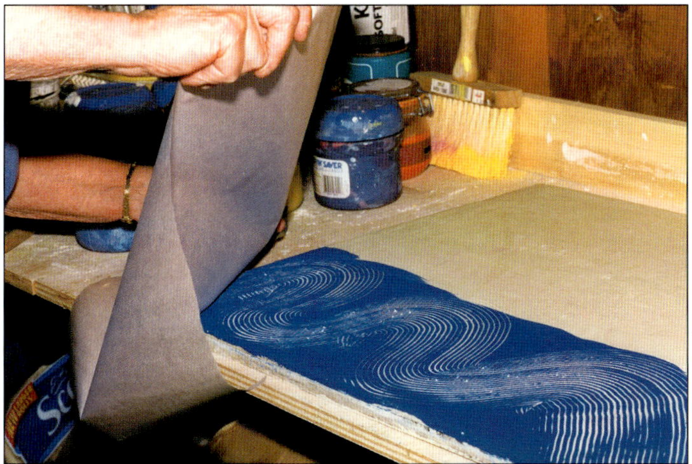

Inlaid design.

Overlap one sheet of newsprint over another. Mix equal parts of blue and red liquid clay in an airtight jar. After thorough mixing, a purple liquid clay will be produced. Apply four to five inch wide thick coats of the purple to both sides of the overlapping sheets. Brush downward from the upper to lower sheet. Allow to dry until the shine disappears.

Apply purple clay.

When the top sheet is removed, a sharp edge will be produced. Spray lightly with water if the clay appears too dry. Invert the wide band of purple adjacent to the textured blue. Cover with clean newsprint and inlay with your roller.

Separate sheets.

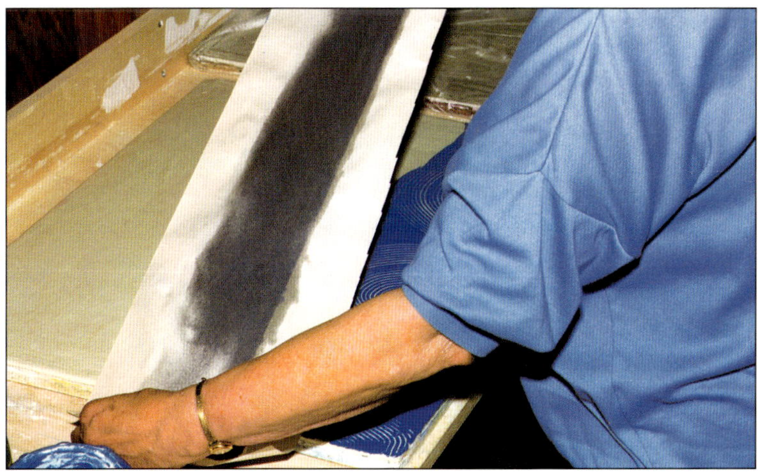

Invert and apply to matrix.

Inlay with roller.

As you remove the newsprint you will notice that the purple is transferred to the clay matrix. If too much purple remains on he newsprint, spray the area lightly and inlay once again. I prefer to have the purple newsprint area somewhat dry so that the color is transferred in a random pattern. The designs and patterns formed are fascinating.

Peel off newsprint.

New design in revealed.

If clay has dried, spray lightly to soften.

Invert the narrow band of purple and inlay close to the top of the slab. Keep in mind that a two-inch border will be created when we finalize the work. It is not necessary to create designs to the edge of the slab.

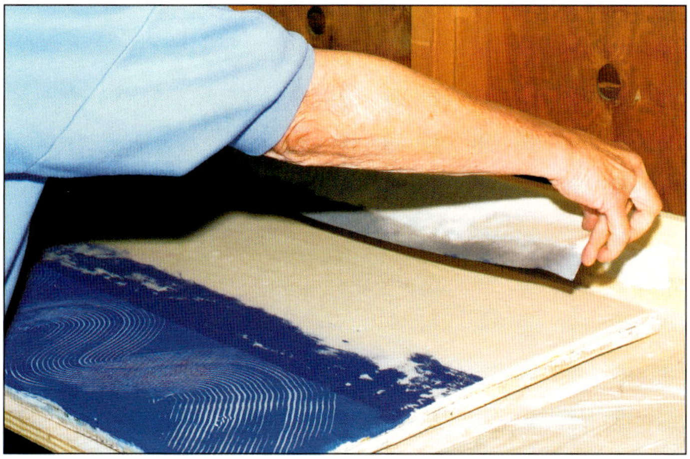

Invert at top of matrix.

Inlay with roller.

Peel newsprint off gently.

When my children were infants, the common ear syringe obtained from our local drug store was part of our household medical equipment. Many years later, this same ear syringe provides unusual artistic patterns.

The technique is simple. To contrast the dark colors of our project with some bright yellow, place some of your thick liquid yellow clay in a separate container and thin it with water. It must be of a thin consistency so that it can be withdrawn and expelled from the syringe without difficulty. Simply suck up some of the thin liquid clay into the ear syringe and drizzle onto a sheet of newsprint. Arrange a series of drops and squiggly lines.

Use ear syringe.

Allow the drizzled patterns to dry until the shine is removed. Be careful! If the clay is too moist, it will smear when inlaid. Spray lightly with water if necessary to soften the clay. Invert onto the purple area and inlay with your roller. Remove the newsprint and note that the pattern has been transferred to the slab.

Be aware that one color of clay can dominate any other when inlaid over it. There is no evidence of muddiness or distortion of color.

Drizzle design.

Allow drying until the shine leaves.

Another approach to design is to use the soft semi solid clay that we had placed in sealed baggies. Take some small pieces of blue and place them on the yellow areas as a contrast. Your esthetic appreciation, and your individual creativity will assist you in their placement.

Press the soft clay into a shape that appeals to you. Dry as needed and inlay with pressure using newsprint and roller.

Invert to matrix.

Apply the soft clay pastel.

Inlay.

Remove newsprint.

Press into shape.

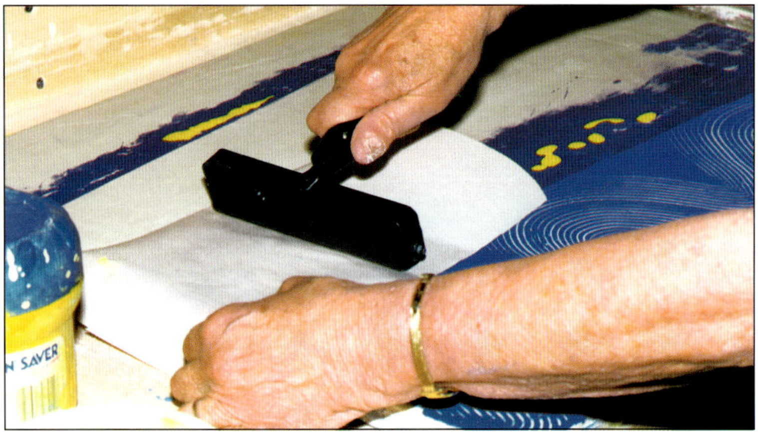

Inlay.

I have decided to introduce some orange-red. Let's use the overlapping newsprint technique with which we produced the blue. A wide and narrow band of the orange-red is our goal.

Orange red.

Introduce orange red

Inlay

In order to increase texture, brush the thick clay in varied directions. This will improve the design. For directional contrast, inlay the orange-red at the upper right corner of the emerging project.

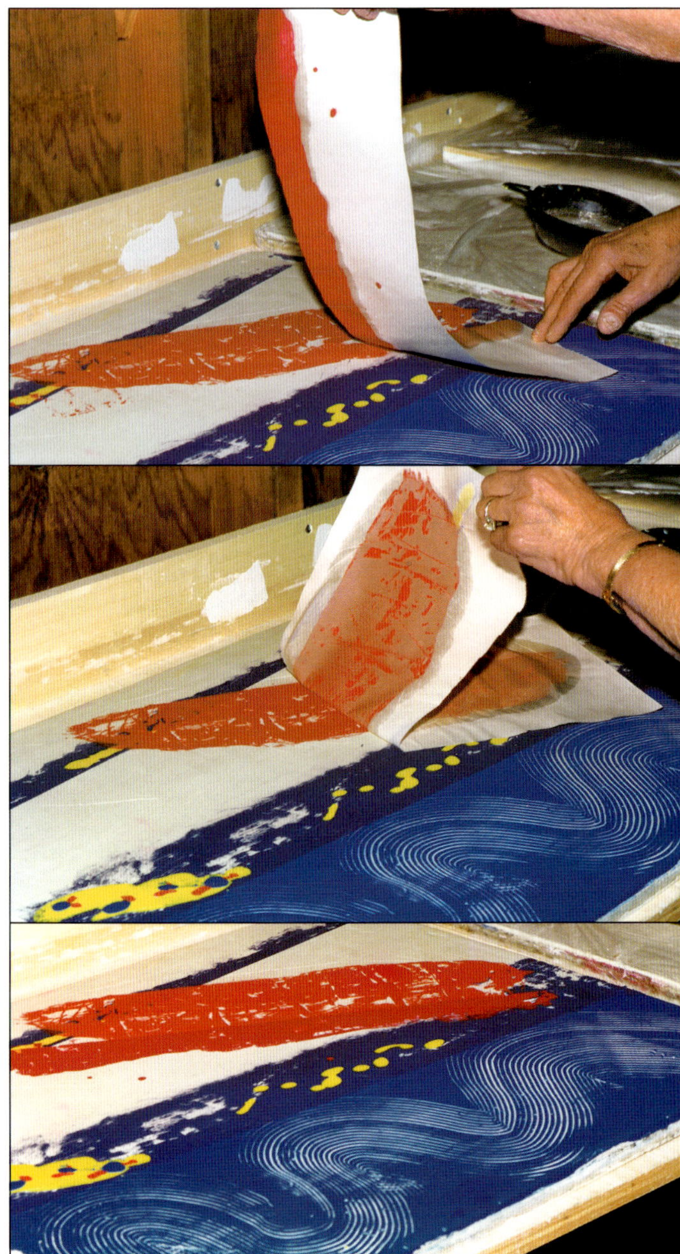

It may appear that we have proceeded with an absence of design. This may be true for I prefer to let my designs evolve as I work. I prefer to accept freedom of expression as planning.

Individual creativity is a personal manifestation of artistic intellect. It cannot be dictated, nor can it be acquired from texts and manuscripts. It can be stimulated by exposure to the arts in all of its forms...visual, performing, and literary. It can be encouraged by proper instruction and encouraged by active participation.

My student seems to be impressed. She will have the opportunity to work at her own creation after I have made her aware of the various tools at hand.

A cardboard tube from a paper dispenser is a handy tool. Dip the end of the tube into some colored clay and press it gently to the slab. In this case, lets us use yellow. A circle of color appears. Continue and form a pattern.

Allow the shine to disappear from the yellow circles and inlay as before using newsprint and your roller.

Dip cardboard tube.

Add circles to pattern.

My thought is to add green to the design. Make it a point to follow your primary impulse. Your first impression is essentially creative. When you second-guess yourself, the creativity is destroyed and reason persists. Reason and artistic creativity do not necessarily gel.

Overlap two sheets of newsprint and apply green liquid clay. After the newsprint removes water from the clay, the shine will disappear. Lift one of the sheets of newsprint as you have done with the blue and orange-red.

Invert the two strips of green and inlay with the roller. You should be experienced with this technique by this time.

Add green.

Add more yellow and blue circles. Finally a streak of yellow in the upper left corner using inverted newspaper technique.

Add yellow and blue circles.

The design is complete. We are ready to pull the impression.

Complete design.

CHAPTER IV

# The Clay Impression

At this point in the process of impressions with clay, it is not unusual for me to feel emotionally drained and artistically depleted. The adventure of discovery has been completed. There seems to be a shutdown of the desire for creative exploration. All that remains to conclude the project appears to be physical rather than aesthetic.

This phenomenon does not relate solely to clay impressions. It occurs when I complete a sculpture and all that remains is the construction of a base.

The sensation is not mine alone. Art, whether visual, performing, or literary, is an emotional process. I am certain that other practitioners would concur. Fortunately, artists possess a propensity to create and the urge for the vision of "Alice in Wonderland" returns. A new project awaits.

Nevertheless, it is imperative that we proceed with the physical production of the impression before the clay matrix becomes too dry. If you must leave the studio, secure the slab within a plastic bag to retain the moisture.

The transfer of the image from the clay matrix to a permanent canvas is quite dramatic. Although somewhat anticlimactic as compared to the creation of the image, it is a revolutionary concept. A press is not required!

The tools are quite simple. You will need a water spray bottle, assorted rollers, a metal tablespoon, some nonwoven canvas, and a roll of dry wall tape, which is used to create borders around the Impression.

Lets do it!

The clay of your matrix image should be slightly moist. It has been continuously inlaid and flattened. Examine the surface and use your fettling knife to remove any foreign material that may have landed on the surface. The presence of any material other than clay will interfere with the transfer.

Spray lightly with water and inlay the entire surface once again. Use a clean sheet of newsprint and a four to six inch roller.

Inlay prior to transfer.

Dry wall tape is a unique product. It actually contains clay on its surface. Tear off a piece of tape somewhat longer than one side of the matrix.

Spray the tape lightly with water and place it on the slab that has been measured. This is your chance to esthetically create your border.

Continue on all sides of the matrix. You now have a bordered image.

Measure length of dry wall tape needed.

Spray tape lightly with water.

Apply the tape on all sides.

The matrix image is now bordered.

We are now ready to produce the image transfer. Remember, this is an impression taken from the matrix and the resultant image will be in reverse form.

Cut a piece of wall liner to fit the size of your slab. Measure it to have an overlap of about one inch. Remember that your dry wall tape produces a two inch border. You can trim if afterward if your wish.

Cut off a piece of wall liner.

Spray both the canvas and the matrix lightly with water. It is important to use the water spray lightly.

Spray the wall liner lightly.

Spray the matrix lightly.

Place the sprayed side of the canvas over the matrix and secure by running your hands over the entire surface.

Place canvas on slab.

With pressure, roll the entire surface with a four-inch roller.

Apply pressure with a roller.

Lift one end of the canvas. See how the image is starting to transfer to the wall liner. The colored clay is being absorbed onto the surface of the canvas. The key at this point is the prudent use of water spray. Excess water will cause the matrix to run and smear. The proper technique is to lift ends of the wall liner, spray lightly to soften the matrix, roll and examine the exposed area. Do this repeatedly by exposing one area at a time.

Once again, lift one end of the canvas and spray with water. Do the same for the other end. Use a tablespoonful to bring pressure at the borders. Similarly, use the tablespoon for areas that resist transfer. A sharply defined image will be produced.

The image starts to transfer.

Apply more pressure at edges using the tablespoon.

Examine your impression by lifting one end of the canvas. The other secured end will keep the work in register. Do not lift the entire print. Alternate each side to expose one-half of the matrix. This is to assure that the impression will always remain in register with the image on the matrix.

Continue to spray and roll, and lift one side then the other to examine.

Compare the transferred image to the slab. When you are satisfied that the intensity of the color on the canvas matches that of the matrix, you may remove the canvas.

Looks good!

You are done. Do not attempt to replace the impression on the slab. It will be impossible to place it back in register.

Remove completely from slab.

Spray with fixative.

Allow the impression to dry for about two hours. The colors will gradually become uniform in intensity.

If desired, you may spray he entire surface with artist fixative for additional; permanence.

What has happened here? Do you recall the test during the introduction to this text? We demonstrated that when moist clay is rolled flat covered by newsprint, a thin layer of clay adheres to the newsprint. We have replaced the basic clay with a vibrant design of colored clay. The image was transferred to an absorbent material as an example of modern art.

During the process, a layer of clay is removed from the matrix slab. Note that clay remains on the matrix slab so that more impressions can be pulled. However, the colors of each succeeding impression will vary in intensity. In fact, the design itself can change slightly depending upon the amount of clay removed by the preceding impression.

The process of impressions with clay actually produces multiple original images that cannot be duplicated. In the past I have successfully produced as many as three impressions from a single matrix. They did resemble each other, yet subtle changes influenced the overall effect. Each was unique in some manner. Each appealed individually to viewers.

If you decide to continue drawing impressions, the second will transfer more rapidly. The matrix will still be sufficiently moist and very little water spraying is needed. Use the dry wall tape borders that remain and repeat the procedure of roller pressure, observation, and spraying if needed. When you are finished simply apply five coats of slip to the slab. Allow the slip to dry between coats. Repeat the process and create a new design.

CHAPTER V

# Exploring the Variables

I have stated that if one hundred impressions were created, there would be one hundred variables. This is quite true. Each stroke of a brush, every installation of color, every texture and undulation of a design is a variable. Realistically, variables lie within the intellect of the artist. They are the form of expression through which the artist delivers his message. They will vary from media to media, from technique to technique, and from artist to artist.

Freedom of expression persisted among artists of the twentieth century. The explosion of thought and technique made them unique within the history of art. I am confident that the twenty-first century will provide yet greater explorations and achievements.

During the creation of our initial impression, a variety of color application techniques were used. A review of the procedures include:

- Direct application with brush to matrix.
- Creation of texture by scraping with a comb-like tool
- Application of color to newsprint and transfer to the matrix by inlay with roller.
- Color drizzled through an ear syringe to newsprint and transferred to the matrix by means of inlay technique.
- Esthetic placement of bits of semisolid or dried colored clay directly to the matrix.
- Color instilled to the matrix in the form of circles by means of a cardboard tube.
- Borders created with dry wall tape.

Lets take a closer look at other techniques of applying color that I have used with success. Rather than doing a complete impression, the focus will be on the individual techniques which will help you examine their potential. Once you are familiar with them, your curiosity and adventurous spirit will lead the way.

The use of stencils and pastels present an interesting variation in technique. The list of stencils include plastic screens, rope netting, paper doilies, leaves from your garden, soda can seals, and even the breakfast cereal Cheerios. The list is unending. If it works, use it!

Stencils.

The method of applying design and color remains the same regardless of the physical nature of the stencil.

• Place the stencil of the dried surface of the matrix.

• Scrape the dried colored clay pastel over your kitchen strainer directly over the stencil.

• Spray lightly with water to set the grated clay.

• Remove the stencil gently. Try not to disturb the register of the stencil. If movement occurs some distortion of design will result.

• Cover the design with clean newsprint and inlay with roller.

Place screen stencil on slab.

Scrape stencil over strainer.

Spray lightly to set.

Lift stencil gently.

Cover with newsprint and inlay.

Completed design inlaid to slab.

Return to another area.

Let's try a paper doily.

A sunburst screen and perhaps some netting.

The sunburst after inlay.

Scrape pastel over netting.

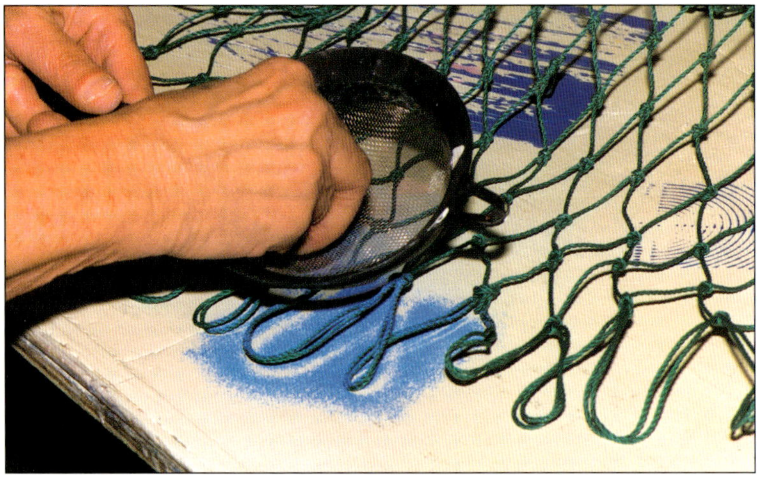

Continue adding colored clay pastel.

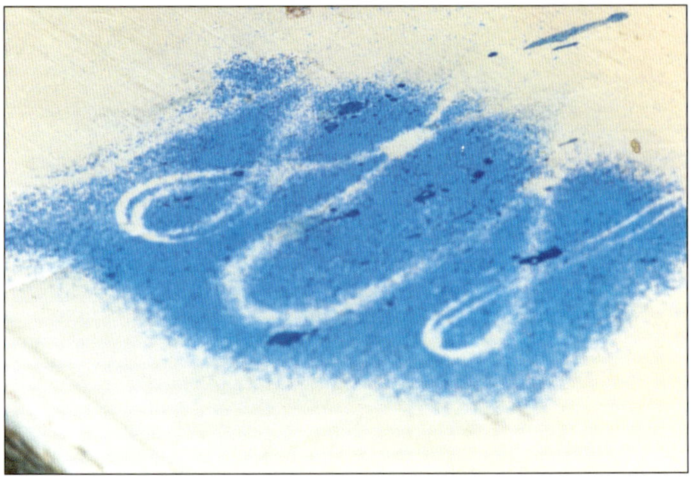

An interesting pattern.

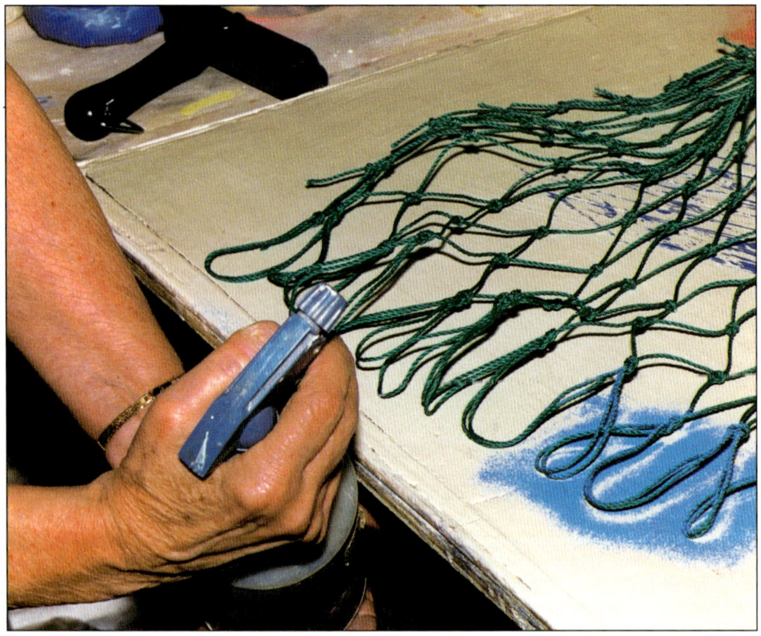

Spray lightly with water to set.

Try a strawberry basket.

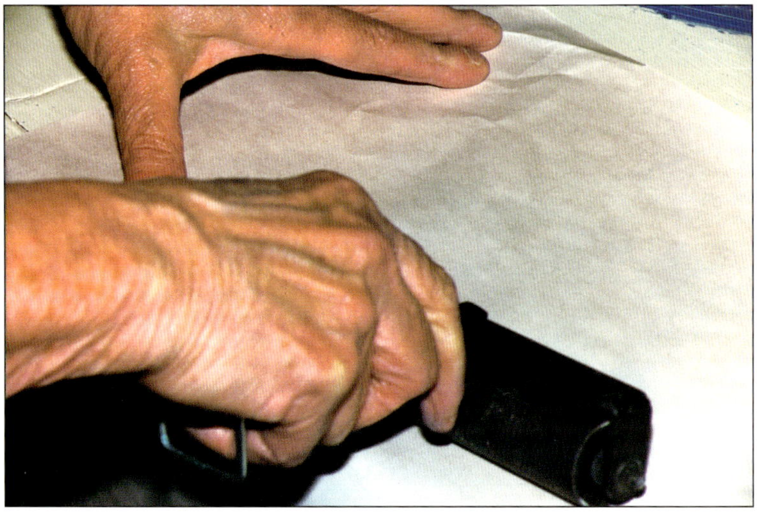

Remove netting and inlay.

Spray to set the clay.

Cover with newsprint and inlay.

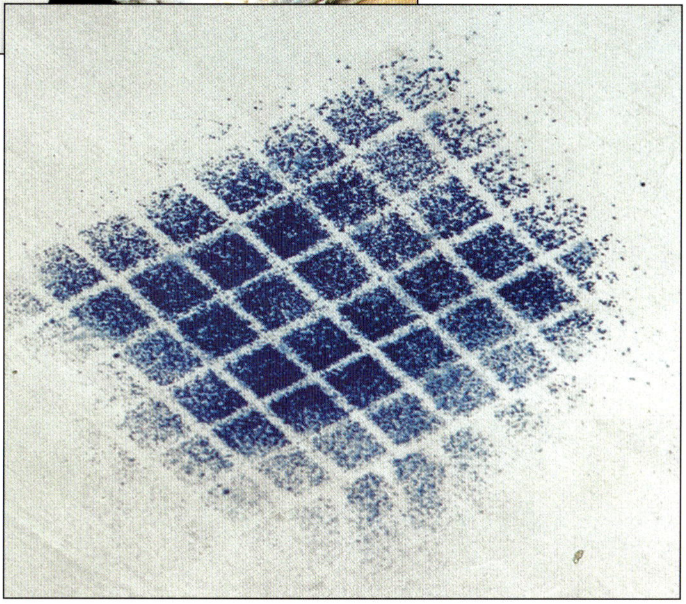

The design.

How about some Cheerios cereal and soda can tabs?

Spray to set. Inlay and remove stencils gently.

Results after inlay. Fantastic!

Note that the inlaid design is flat.

Texture applications are created either by direct application to the matrix or by transfer to the canvas from newsprint. These are techniques that we have investigated. You should be familiar with them by now. The tools are simple everyday household appliances. The photo below illustrates some that I use. Pictured are heavy stiff paintbrushes, cake decorators, a wire barbecue grill brush, brayers, hair rollers, and paint scrapers. Anything goes! It will work for you. Use them!

Texture tools.

Let's try some of the various tools that I have acquired.

A scraper that I found at a flea market has produced interesting work. In this example, I have applied a heavy coat of liquid clay directly to the matrix. The texture is obtained by scraping through the coating. This is followed by the routine technique of inlay using newsprint and roller. Remember to allow the texture to dry slightly before the inlay. If the texture is excessively moist, it will smear.

Applying color to the matrix.

Scraping with a tool.

A brayer will do a similar job. Roll the implement through a heavy coat of colored clay that has been applied to newsprint. The brayer is then rolled lightly over the clay matrix to create the design. Inlay as usual.

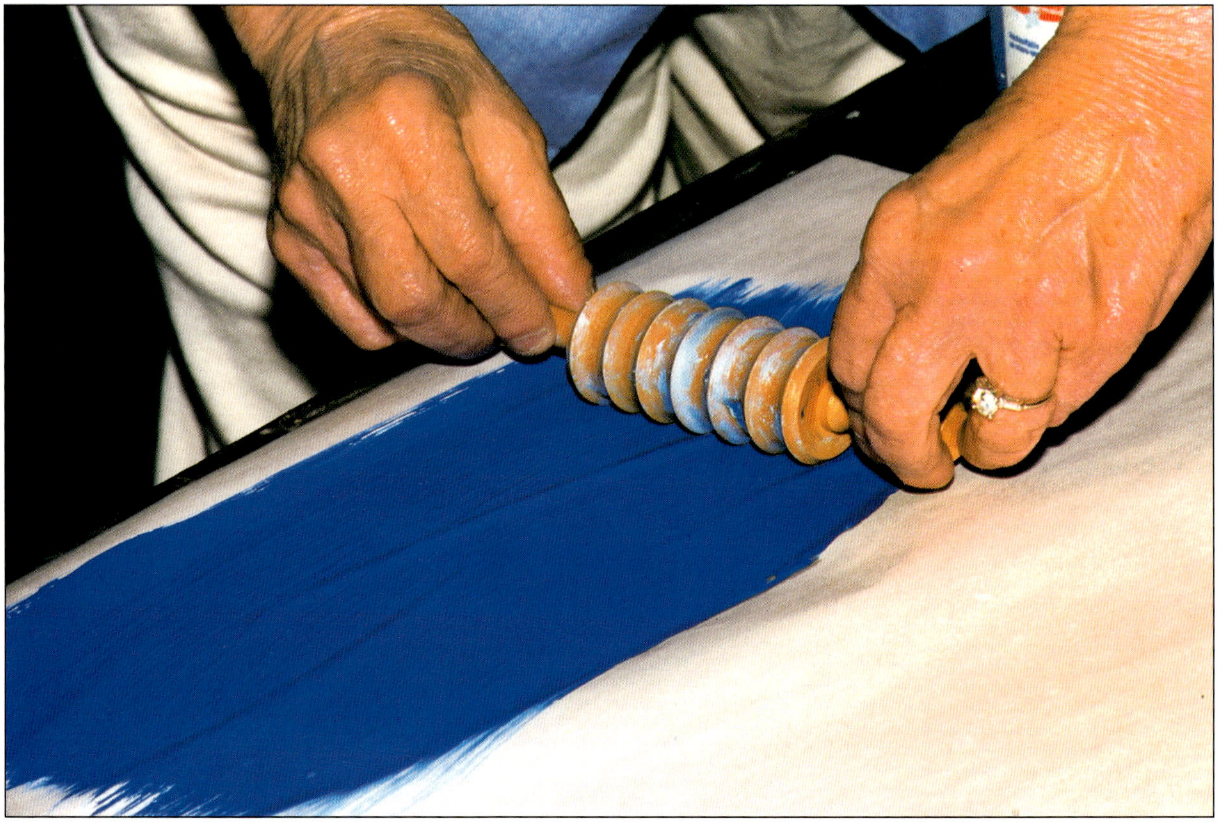

A brayer will do a similar job.

Run a heavy bristle brush through a heavy coat of liquid clay. Make sure the bristles pick up enough of the clay to be brushed on the matrix. Inlay as usual.

Use a heavy bristle brush.

After inlay.

The availability of stencils and texture tools vary with the individual artist. You will find sources of patterns in your every day shopping or browsing experiences. Keep your eyes and ears alert. There are innumerable variations of these procedures that await your individual creative expression.

Examine the impressions in the gallery section, Chapter VI, pages 68 to 82. Notice the use of texture and stencil technique throughout the series. This is the magic of *Impressions With Clay.*

CHAPTER VI

# Gallery of Clay Impressions

### Clay Impressions and Photographs by Milt Liebson

"Aquarius" by the author, clay impression, 20" x 24". 1999.

"Desert Flora" by the author, clay impression, 20" x 24". 1999.

"Mood Indigo" by the author, clay impression, 20" x 24". 1999.

"Painted Desert" by the author, clay impression, 20" x 24". 2000.

"Interlude" by the author, clay impression, 20" x 24". 2000.

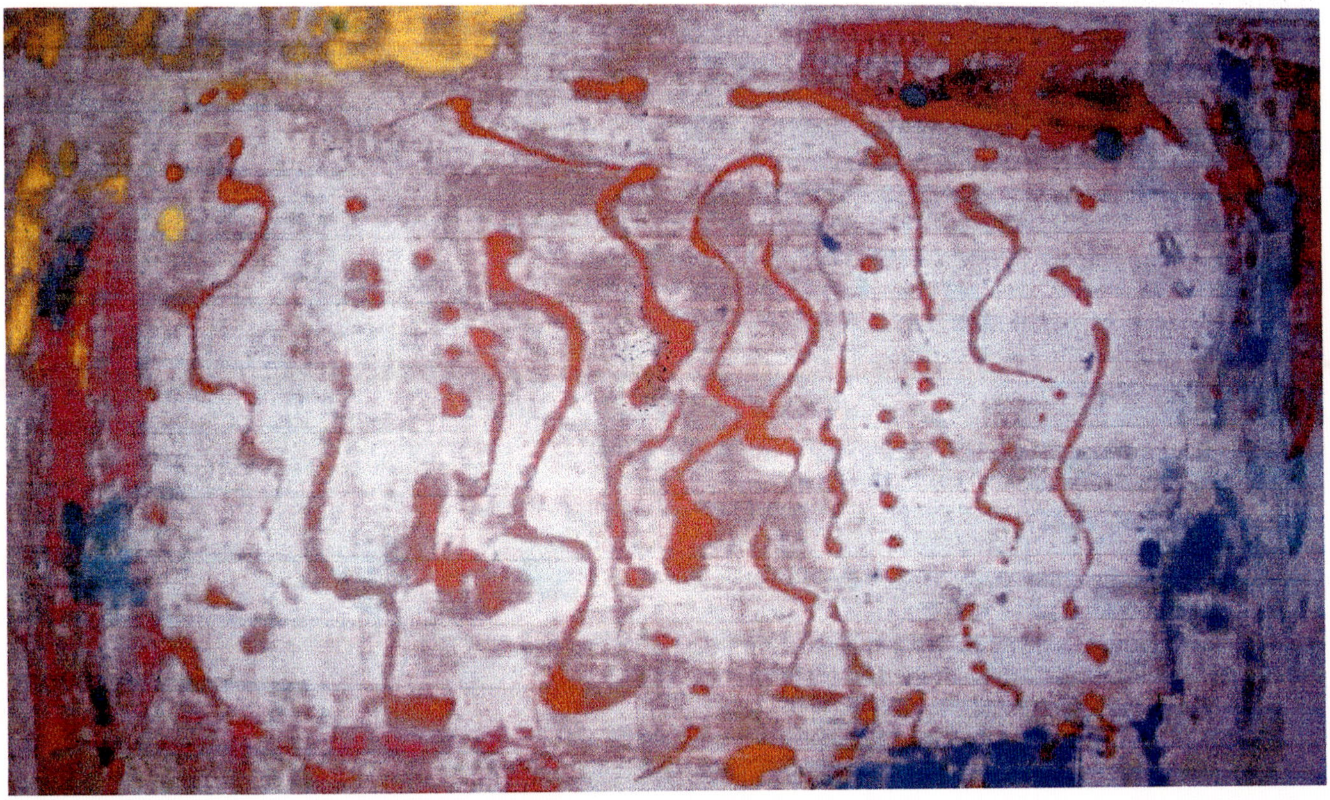

"Celestial City" by the author, clay impression, 24" x 36". 2000.

"Kaleidoscope" by the author, clay impression, 20" x 24". 2000.

"Game Board" by the author, clay impression, 20" x 24". 2001.

"Equinox" by the author, clay impression, 22" x 28", 2001.

"Harbor Lights" by the author, clay impression, 20" x 24". 2001.

"Indian Summer" by the author, clay impression, 20" x 24". 2002.

"Sunspots" by the author, clay impression, 20" x 24". 2002.

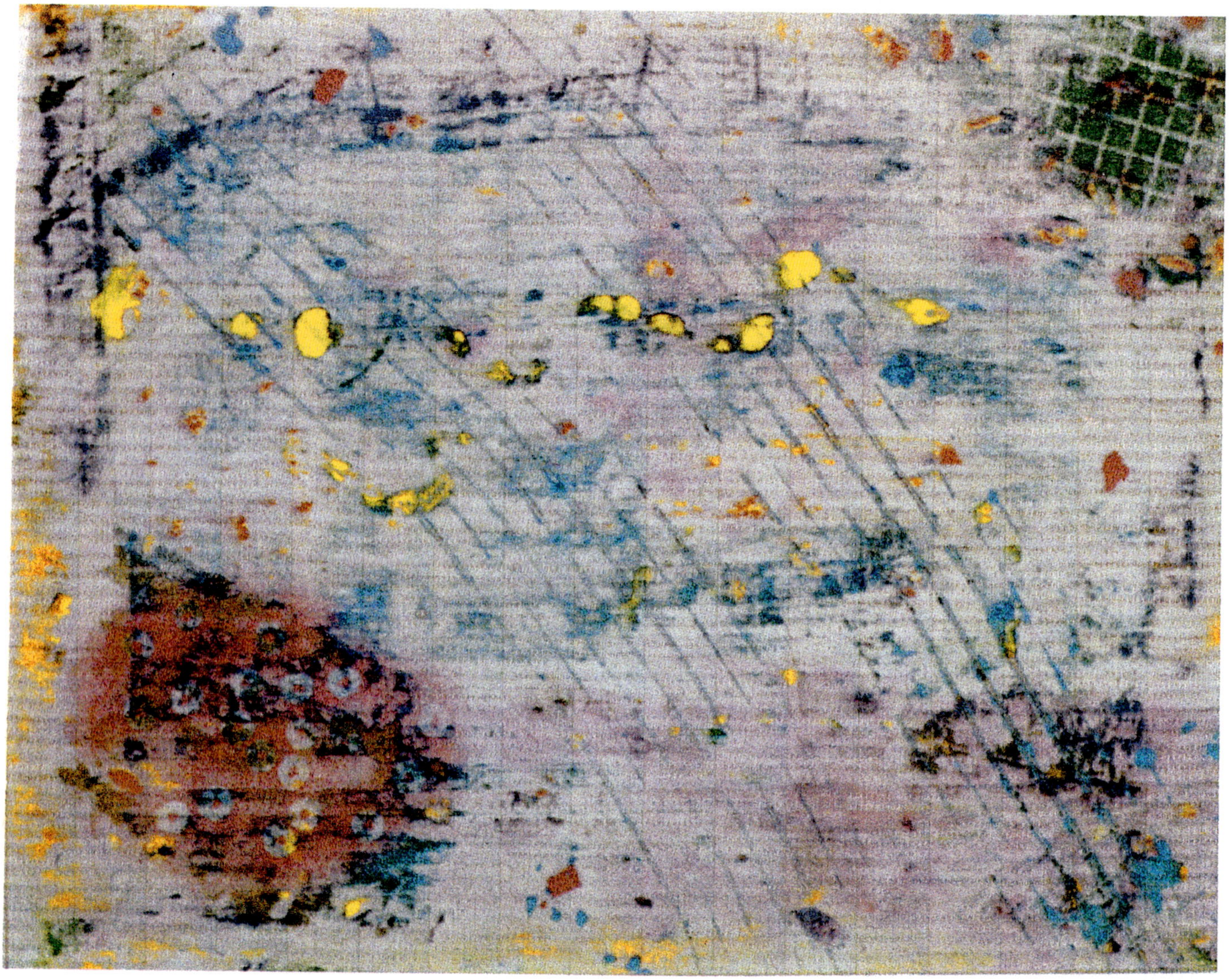

“Visions” by the author, clay impression, 20” x 24”. 2002.

"Beach Party" by the author, clay impression, 24" x 36". 2003.

“Field of Dreams” by the author, clay impression, 20” x 24”. 2003.

"Autumn" by the author, clay impression, 20" x 24". 2003.

"Metropolis" by the author, clay impression, 20" x 24". 2003.

"Inner City" by the author, clay impression, 24" x 36". 2003.

CHAPTER VII

# Lila Creates an Impression with Clay

We have come a long way, Lila. You have been straining at the bit. I can sense your impatience. Go for it! Create an impression! I'll stand by and offer constructive criticism if needed. Take over.

"Thanks, Milt. The hesitancy that I had experienced at the inception of the procedure has faded away. I want to do this."

"I know that I can use the same slab for new work. However, it does seem a bit overloaded with clay from previous Impressions. I'm going to use my three inch board to scrape across the slab to remove some of this excess."

"Gosh, Milt, as I remove some of the excess clay, dramatic visual effects are produced by the appearance of previously inlayed images."

That is quite right, Lila. As a matter of fact, "Visions," in the Gallery was created with this technique and a little added emphasis. This is the drama of the process ... exploration, discovery, and innovation.

"The excess clay has been removed. I will apply five coats of white slip and allow drying between the coats. I know that I will need a surface that is dry to the touch after the final coat."

"I'm going to scrape my pastels across the strainer to create design. I will start with pink, then yellow, and some green. This will be followed by a light spray with water so set the shavings."

A dry surface.

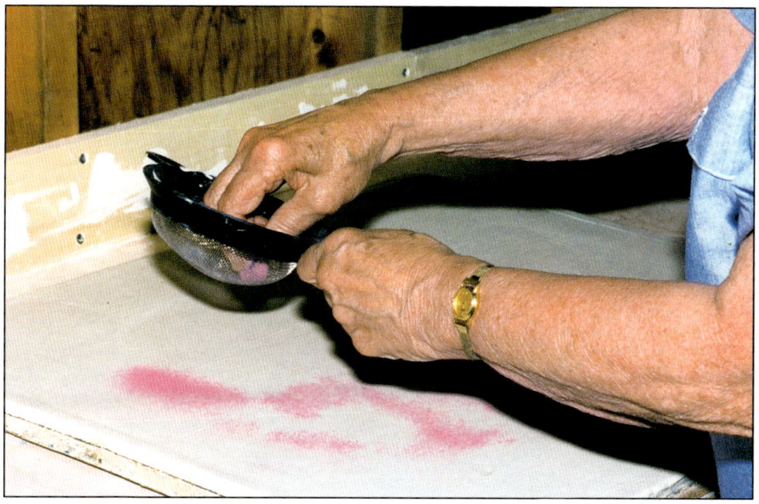

Scrape pastels.

Pink, yellow, and green.

Spray to set.

"Before continuing, I will inlay the design by using newsprint and roller that you have demonstrated. Then I'm going to add a dash of green by applying it directly to the slab with a wide brush. A sharp pointed utensil drawn through the moist liquid clay will create texture."

Add a dash of green.

"I like the dramatic touches instilled by drizzling with the ear syringe."

Drizzling with ear syringe.

"And more contrasting design using blue applied directly to the slab with a wide brush."

More contrasting design.

Lila, it is good technique to inlay frequently. If the clay is too moist, it may smear when rolled with newsprint. Allow the design to dry until it is a little firm. Spray lightly with water if it becomes overly dry, then inlay

"Thank you, Milt. I got carried away with creating the design. I'll use newsprint and rollers to inlay."

Newsprint and rollers.

"My supply of yellow seems a little thick for use with the ear syringe. I recall that it should be fairly thin…almost watery before it can be drizzled to newsprint or the slab. I'll just add some water to a small amount and mix well before using."

Thin with water.

Drizzle.

"A fill in of color using pastels and the strainer."

A fill-in of color.

"And a final inlay of the entire slab with newsprint and roller."

A final inlay.

" I'm using the lightly sprayed dry wall tape to create my border. The excitement and anticipation is growing."

Applying wall liner to slab.

Lightly sprayed dry wall tape.

Added pressure.

Continue applying border.

"The transfer of the design is increasing with each added pressure. I'm pulling the Impression. This is just amazing. I love it!"

"I have cut a piece of wall liner to fit the slab. I recall your direction to control the use of water spray at this point. Spray lightly, roll with pressure. I'm going to use the edge of a spoon as you directed. I will continue with pressure until the Impression starts to form.

I can see the design starting to transfer.

Lila and her completed clay impression. I would call this excellent for a first timer.

Lila, I'm proud of you. You have performed admirably. Allow the work to dry for about two hours. Spray with artist fixative if you wish.

## CHAPTER VII

# Matting and Framing

We have explored the technique of impressions with clay. Quality work has been produced. Are our efforts completed? I think not. There is the question of a proper vehicle for presentation. This has to mean matting and framing.

There is no doubt that for the sake of quality and professionalism, the most efficient procedure is to take your work to a qualified framer for completion. Their expertise and experience in frame and mat selection would enhance the beauty of your work and encourage sales.

This thinking would be suitable for a onetime effort or for limited output. However, what about the serious artist and professional? Undoubtedly, completed work must be properly represented for presentation to buyers, exhibitors, and curators of galleries. A prolific amount of work produced would make the expense of a professional framer impractical. Furthermore, many collectors of art prefer to deal with a framing professional to coordinate with individual preferences. In the latter case, premature expensive framing by the artist would be wasted.

The approach that I follow is to presume that buyers will want to have the purchased work framed professionally. I would be unhappy to see my investment in exotic framing discarded. Therefore, I will proceed with the simplest and most cost effective method. That approach would be to do the matting and framing myself.

For the professional artist or serious experienced hobbyist the matter of doing our matting and framing is "old hat." It is a procedure that is part of art training.

Allow me to direct my attention to the novice and to those who have not had experience in this area. The procedure is not difficult and with a little practice, you can perform as an expert. There are several options:

- Shrink wrapped with plastic film over foam board mounting with or without matting.
- Fastened to wood stretch frames and placed within a suitable frame with or without matting and glass.
- Under glass with or without matting using metal section frames.

It is critical that all materials aside from the frame and glass be "acid free." I am referring to the shrink-wrapped film, mats, backing board, backing paper, and adhesive tape used in the framing process. Materials that are not acid free will discolor in time, and may discolor your artwork. Be certain that the materials you buy indicate that designation. Do not minimize this requirement.

Shrink-wrapping allows you to seal an impression that has been mounted on a foam board. This places your work in a dirt, dust, and moisture-resistant environment. It serves as an inexpensive means of presenting your product to buyers and curators prior to framing. Work, which has been shrink-wrapped, is rarely hung for display. Rather they are catalogued at galleries within a folder of unframed art for sale at prices excluding framing. It is at best a temporary procedure.

A further deterrent is that the cost of the plastic film and heat sealing system is far too expensive to purchase for infrequent use. Professional framers can provide the service at a fairly modest cost.

If your preference is to fasten the work over wooden stretch frames as you would conventional artist canvas, be sure to leave two inches in addition to the border on all sides of your printing material. Use a staple gun or steel tacks to fasten over and to the back of the wood frame. Use canvas pliers to pull the material taught to the opposite side and fasten to the back of the frame. Repeat for the other sides. Fold and staple down the corners. The stretched canvas can now be placed in a decorative frame with or without matting or glass.

For my work, I prefer option three. I routinely mount all of my work under glass or plexi-glass, with matting, within metal frames. This type of framing provides an attractive cost effective means of displaying my work. My price structure includes the impression and the complete frame. In almost every instance the art and frame are sold intact.

In some cases, I have agreed to an adjustment in price to exclude the frame. These instances are rare. Most buyers prefer to hang their acquisition without delay. Changes in framing may occur after the collector has time to "live" with their possession. Any changes become the obligation of the owner of the art.

Be sure to indicate a title for the work in the lower left corner of the border and your signature in the lower right corner. This establishes your copyright. If you provide matting for the work that covers the borders, do the same for the mat. In this matter, your copyright will be protected if the mat is removed.

To apply the signatures, I recommend the Eberhard Faber "Design Ebony" pencil. The writing will be jet black and extra smooth in application.

Lets talk about the mat. It is a critical part of your framing. You can frame without a mat and mount the work on a suitably colored backing board. The white border of the Impression against the color of the backing board will be quite effective. That is your option. I prefer the mat.

Additionally you may use the cut out portion of the mat to expose a portion of the border of the work. The signature on the original will be exposed and additional writing on the mat would not be necessary. It is a matter of personal choice. I prefer to use the bevel of the mat to outline the design.

A striking clay impression accentuates the sophisticated, charming, and somewhat soft elegance of a fiber mat. Deep bevel adds depth and beauty to my art work to be framed. It will bring your impression to a higher level of acceptance.

The thickness accorded by a mat frame serves a definite purpose. It provides a non-contact area which separates the artwork from the glass or plastic placed over it as part of the frame. This space is particularly useful since a circulation of air is needed between the art and the covering. This is required to prevent condensation of moisture and subsequent formation of mildew and mold, or buckling of the matting and or the art itself.

It has been shown that paper will expand and contract with changes in humidity. This space within the mat must be provided within the mat frame to allow proper air circulation. This will extend the life expectancy of the art.

For the novice or artist of limited output, the easy way, though not the cheapest, is to use beveled ready-made mats. There are several brands. I have always been satisfied with the "Pyramid" and the "Crescent" brand. These are available at local art supply stores. Certain mail order vendors offer quantity discounts of about fifty percent off full retail prices.

Quality ready-made mats provide you with a totally acid-free environment for your art. They feature acid-free core, acid-free backing paper, and acid-free board. The mats are supplied in twelve artist sizes, three styles, and fifteen colors. All that needs to be done is to mount your work under the beveled top layer, and place in your frame. It is simple and efficient.

The down side is that the range of sizes of ready-made mats may not satisfy your needs. Because of this, and the monetary savings, the serious artist will prefer to cut their own mats. I do.

Mats, backing paper, and backing board are usually available in sheets of thirty-two inches by forty inches. Additionally, these materials can be ordered in a forty-inch by sixty-inch size.

There are a great variety of mat cutters available to you. A glance at any supply catalog will reveal such names as "Designer's Edge," "Logan," "Alto," and "Fletcher" among others. They are all good.

The range in prices within each manufacturer varies with the size of the cuts to be made and the refinements available. A low priced cutter may be limited to a thirty inch length cut whereas a sophisticated model may offer cut length up to sixty inches. Retail prices range from under one hundred to over seven hundred dollars.

For my use, I employ a Logan Simplex 40" model #700SMG. This model falls within the lower half of the price range for mat cutter. It provides me with all the advantages that I require. I do recommend this model. Shop around for price. Be sure to check the mail order art suppliers. I made my best deal with one of them.

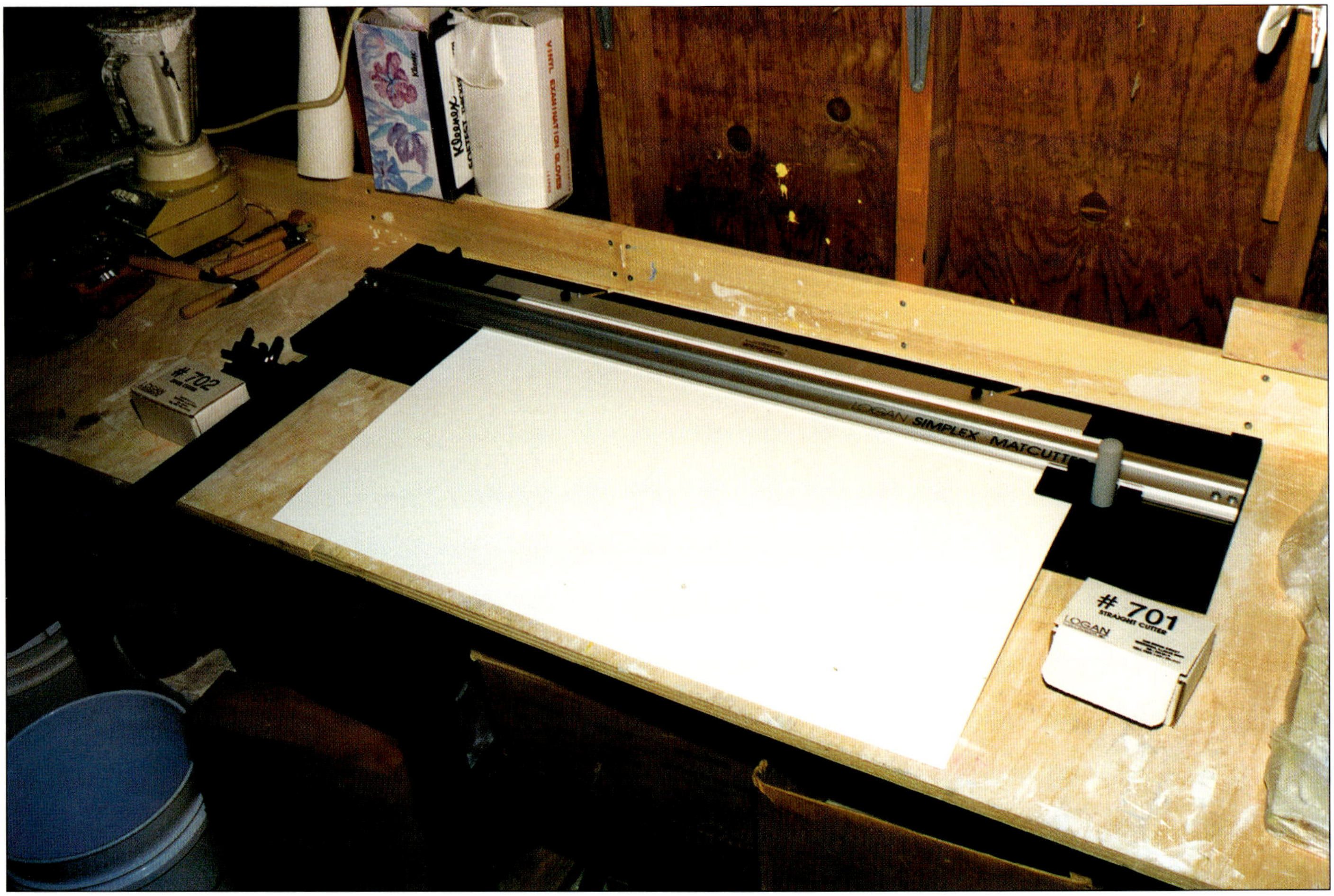


Logan Simplex mat cutter, model 700 SMG

Cutting a mat is not difficult. You might even discover it to be fun! The instructions that accompany each mat cutter adequately describe the procedure.

The width of a mat frame border is really determined by what is most pleasing to you. Do not hesitate to produce wide borders. Small, narrow borders tend to distract from the patterns of the art. Symmetrical borders provide a sense of balance for most artwork. A bottom border being fifty percent wider than the top and sides is always acceptable. Such a mat design will help to establish balance and produce a bottom-weighted feel.

Several experienced framers have released books and videos to illustrate the assorted styles of mat cuts. You will become familiar with single mats, double mats, offset mats, v-grooves, decorative mats, and variables of these. It is part of the adventure to experiment by mixing colors, shapes, and configurations. You will find that with a small amount of material and a few basic tools, you will be able to create attractive mats with the best of them.

EPILOGUE

# The Art World Awaits You

What shall we do with our impressions with clay? Do we file them within folders or hidden archives awaiting acceptance by friends and family? To do so would be to violate the basic concepts that have been the foundation for creative arts for centuries.

You have been exposed to a new and dynamic art form. Bring it to the attention of the art consuming public. Recognition and acclaim is awaiting you.

Art must be shown. The art form is crucial for the maintenance of equilibrium in our civilization and your participation is vital. Art is the means by which future generations will learn of our present endeavors, as we have learned of the past. Without art there is no civilization. What remains is chaos.

There is always the potential for profit. For today's visual artist this potential is highly suspect, but not beyond reach. The separation between professional and nonprofessional is not financial reward alone. I prefer to think that it is the attitude of the individual toward the perfection of his or her talent, the increased recognition of inherent ability, and its contribution toward he stability of our world.

# Index

# Addendum

To those who have enjoyed this book, it is important that I recognize the encouragement and cooperation afforded by my "student," Lila.

Lila was my wife of almost sixty years. Initially meeting as freshmen in college, a lifetime of happiness, love, and affection ended on September 25, 2003. That is when Lila passed away prior to the publication of this book.

I shall miss her with all my heart.

Milt Liebson